The Fantastic Book of Physics Jokes:

For Everyone, Not Just Physicists

UNCONVENTIONAL
PUBLISHING

www.unconventionalpublishing.com.au

Shane Van

Shane Van

ISBN – 978-0-6452206-9-8 HARDCOVER

DISCLAIMER: THIS BOOK IS A WORK OF FICTION AND IS
NOT TO BE TAKEN SERIOUSLY; THERE IS NOTHING IN THIS
BOOK THAT SHOULD BE TAKEN AS FACT, ESPECIALLY
SCIENTIFIC FACT.

Table of Contents

Preface

Unconventional Publishing proudly present a book that will make you laugh and think of science and physics differently. Physics is probably the most popular of sciences, just look at many of the famous scientists, the majority being physicists. In sci-fi movies, Who has the answers? the physicists, unless there is a geologist present. Nothing in this book should be taken as fact and is only here to put a smile on your face; no part of this book is meant to offend. If it does in any way offend you, you can always put it down and move on to something else.

This book is the third in a series of books and follows on from *The Fantastic Book of Chemistry Jokes* and *The Fantastic Book of Biology Jokes.* This series roasts all sciences and everyday professions. The series will cover professions from chemists to physicists and doctors to plumbers.

Electricity

I can tell you some facts about electricity and it would really shock you

I was at an open-air symphony the other day and the place was struck by lightning. Luckily only the conductor was hit

What is the name of the detective who solves electrical crimes?
Sherlock Ohms

There once was a train driver in Bulgaria who was horrible at his job, turning up to work hung over, falling asleep at the wheel, missing red lights and train stations. One day he falls asleep, misses a red light, and ends up de-railing the train. A lot of people were injured or killed in this incident. The train diver was arrested and found guilty of manslaughter. He was then sentenced to death. For his final meal, he just wanted a punnet of strawberries. He was still munching on them when he was strapped into the electric chair. The executioner threw the switch, sending sparks everywhere causing the room to black out. When the lights came back on, the train diver was there still alive and smiling.

Due to an old Bulgarian law if someone survives an execution then it is an act of God and he is then set free. So, the train driver is set free and gets his old job back. Sure enough, he doesn't learn his lesson and once again misses a red light, derails the whole train, and ends up killing people. He is then arrested again and sentenced to death a second time.

For his final meal, again, he just wants a punnet of strawberries. He walks up and sits in the chair and is still munching on the strawberries as they strap him in. The executioner once again throws the switch, sparks fly, and the room goes black again. When the lights turn back on, the train driver is sitting their smiling again. Due to the Bulgarian law he is set free and goes back to work.

While back at work again, drunk, the train driver this time ends up crashing into another train, derailing both trains and killing even more people. He is arrested a third time and sentenced to death a third time.

For his final meal he requests another punnet of strawberries.

"No way in hell are you getting more strawberries!!! This time you will definitely fry!" said the executioner.

"You seem to be confused," says the train driver, "The strawberries have nothing to with it. It's just that I am a bad conductor."

Electrical engineers like to keep their news current

Did you hear about a guy that was siphoning electricity from a museum?

He was a joule thief

Is an electron pessimistic or optimistic?

They are pessimistic because they are always negative

What did little Jimmy's mum do when she caught Jimmy zapping his sister with static electricity?

She grounded him

I bought a woollen sweater, but it kept shocking me from the static electricity. The shop gave me another one free of charge.

If electricity always follows the path of least resistance, why doesn't lightening always strike in France?

My teacher was busy explaining electricity, and I was like…. "Watt?"

What did socialist countries use before candles?

Electricity

A dangerous electrical surge walks into a bar.

The barman asks, "Why the long phase?"

You're a joule per second Harry
I'm a watt??

A friend of mine recently just graduated from university, he became an electrical engineer. He was a star pupil and aced all his subjects. His first job was for a wealthy man that wanted to put up a huge electric fence right around his property. On the last day of the job, my friend turned it on, then tripped and fell on to the fence. I am not sure if he's dead or is still in his current job.

Why can't you take electricity to a party?
Because it doesn't know how to conduct itself

Why do electrons carry to identify cards?
Because of their wave-particle duality

I tired to make a belt of joules, it was a waste of energy

A lethargic eunuch admits himself into a hospital.
"Doctor, I think I am suffering from fatigue I've lost all my energy; I feel sleepy all the time."
The doctor runs a series of tests, checks his bloods, goes over his diet but can't seem to find the cause of the eunuch's condition.

"I'm sorry, but we just don't know what is causing your fatigue. We've tested your iron and sugar levels, you're not anaemic, you don't suffer from depression, you have no sleeping disorders, we've tracked your diet. We are stumped."

A physicist who was in the bed next door and overheard the conversation, interrupts.

"Sorry, I didn't mean to eavesdrop, but I know exactly what is wrong with him."

"How on earth could you know what's wrong with him!" complains the doctor.

"Well, you see someone, has stolen his joules." says the physicist

A young student was asked to do a report on Nikola Tesla. The student spent all night studying and working. He came to school the next day and went to hand it to his teacher.

"So where is it?" asks the teacher.

Looking through his bag, trying to think of an excuse.

"I'm sorry Miss, my dog at my ohmwork"

Fundamental Forces

Gravity is a very important part of physics, if you remove it, you are only left with gravy

I was reading a book on anti-gravity, I found it very uplifting

Your parents are so fat that when they have sex, they radiate gravitational waves

A farmer put shoes on his horse, but when he put the horse in a field it couldn't move anywhere. It was the magnetic field.

Why do people hate gravity?
Because it is always dragging people down

What did a magnet say to the another magnet?
Seeing you face on I thought you were repulsive but watching you from behind I find you attractive

When life gives you lemons, make lemonade. If life gives you apples, then make a gravitational formula.

Why are astronauts always happy in space?

Because gravity can't drag them down

Why did the toilet roll down the hill?
To get to the bottom

Was that cocktail a magnet?
Because you just got attractive

What do you call a criminal that stole a Tesla?
A joule thief

What to do they call Teslas in Germany?
A Volts-wagon

Chuck Norris is the force

Where does bad light end up?
In a prism

A guard walks past a jail cell. Inside there is a lonely
photon just leaning against the wall.
The guard asks, "What are you in for?"
The photon replies "Forbidden transmissions."

*How many physicists' does it take to change one light
bulb?*

10, 1 to change the light bulb and another 9 to co-write the paper.

I love the way the earth rotates, it just makes my day

Did you hear about the photon that walked into a bar?
Just kidding photons can't walk

How many physicists does it take to change a light bulb?
Only 1 but he'll replace it 3 times, measure the lumens, plot a straight line, calculate errors then extrapolate to zero concentration

A photon checks into the Hilton one night, when the bell boy asks, "Sir can I take your bags?"
"I don't have any, I always travel light."

How many Theory of Relativity supporters does it take change a light globe?
Two. One to hold the light bulb while the other spins the space around the light bulb.

Everyone thinks photons are the friendliest particles. It's because they keep waving

Warning signs at physics labs;

"Do not stare into laser with remaining good eye."

Why did two photons have an argument?
It was because of the interference

Why can't Catholics travel at light speed?
Because they have mass

What do scientists enjoy the most at sports games?
The wave

There is a secret party for physicists. Each year around the globe they meet at a secret beach, they call it the popular wave function

How can you find Doppler's car in the car park?
It has a red sticker saying. "If this sticker is blue you are driving too fast."

Space

On Amazon, our solar system has only one single review, it is a 1 star

Astronomers who study the sun have a flare for research

The moon was mad at the earth, but it was only a phase

What was the name of the first satellite to orbit the earth?
The moon

If a someone is trying his best to give you the moon and the stars, why not reward them with Uranus.

Solar neutrinos are penetrating you every second, can I join?

Most rocket scientist are actually skilled archers, they are good at arrow dynamics
Did you hear about the claustrophobic astronomer?
He just needed a little space

Hear about the astronomer that was throwing a party?
He needed help to planet

An astrophysicist's girlfriend said she needed some space and time, so he gave her a calculator to work out the speed.

Astronomers in the 1500's: We are the centre of the Universe
The Sun: Hmm, well I have some bad news for you.
The Milky Way: Yeeeahhhh, um, hate to say something but…
The Rest of the Universe: "Ditto."

Hear about the astrophysicist that wanted to put an observatory in his house?
The cost was astronomical

What does an astrophysicist and a proctologist have in common?
Every time they study Uranus, it's always lying on its side.

What is the difference between an astronomer and astrologer?
Only about 60 IQ points

Why is The Big Bang, the first original hipster?
It describes things before it was cool

What do physicists call an orgy?
The Big Bang

What came before The Big Bang?
The Big Foreplay

What do people and the universe have in common?
They both start with a big bang

Your momma's so fat, her fart caused the Big Bang
A lot of people have been saying The Big Bang Theory
disproves God.
I mean, it's a funny show and I like it, but I wouldn't say
that.

Your momma's so fat, her toenails don't have nail polish
on them. They are just red shifted

Well, if the Big Bang happened 13.8 billion years ago, and matter cannot be created or destroyed. All of our bodies are made of matter, that is actually 13.8 billion years old. Sure, it might have changed densities due to stars going supernova and undergoing nuclear fusion caused larger and larger molecules to be created, but it is still 13.8 million years old. That is why you honour, I never asked her for her ID.

Tell a man that there are at least two trillion galaxies out there and he will believe you, tell that same man that the bench still has wet paint on it, and he will go and touch it.

A young couple both studying astronomy are trying to relate the topic to their love life.
"Honey, your penis reminds me of a star."
"Really," he says, "is it because it makes you hot and brightens your day?"
"No, it's because it's a white dwarf."

What did Mars say to Saturn?
Give me a ring sometime

The Mars rover suddenly stopped transmitting singles back to earth, so NASA decide to bring it back. It cost millions and was a really difficult task. After well over a year, it was finally back home on planet Earth, and all the scientists were eagerly gathered around it, patiently waiting to turn it back on to reveal the missing data. It exploded and sent debris everywhere. Luckily there was only one casualty, Sylvester the office feline and employed rat catcher.

Moral of the Story: Curiosity killed the cat

Aboard the ISS the newest crew member has decided to make a morning coffee for everyone. He spends at least 20 minutes looking for the condensed milk.

"Hey, do you know where the milk is?" he asked

"Sorry we can't use milk in space, no one can, here, use cream."

Why did the sun go to school?

To become brighter

Did you hear about the restaurant that NASA opened on the Moon?

Great food, but no atmosphere

What is an astronomical unit?
An absolutely huge apartment

Why did the black hole fail high school?
He wasn't bright enough

Why doesn't a physicist wear black socks?
They are afraid of creating black holes

I failed my physics exam today; the question was, 'Who discovered the first black hole?'
Apparently, Ron Jeremy wasn't the correct answer

If you weren't impressed with the images of the black hole, you clearly didn't understand the gravity of the situation

Why are black hole jokes so bad?
Because they suck the most.

Why doesn't the sun need to go to university?
He is already has a few thousand degrees.

Space kept running around in circles, jumping through the air, always talking and never sitting down. It was hyperspace.

Since I was little, I always wanted to visit the Air and Space Museum. I was so upset when I got there, it was just an empty room.

Sherlock Holmes and Watson are out camping in the middle of the woods, when Sherlock bumps Watson awake.

"Tell me Watson, what do you see?" Sherlock asks him

"I see stars, thousands of stars, I see the moon and I see a faint cloud which is the arm of the milky way." Watson answers.

"And what do you deduce from that?"

"Well, that we are infinitely small in this cosmic ballet that has been going on for millions of years. The moon evolves around us, given host to a whole lot of nocturnal flora and fauna that depend on it. The thousands of stars we can see are only a speck of the galaxies which contain millions of stars with planets on their own and some of this must have some life on them. What about you Sherlock?"

"Hmmm," he says in a thoughtful tone, "I believe someone has stolen our tent."

So, let me get this right, an astrophysicist can deduce a stars composition speed, size and direction from radio signals billions of light years away, but I can't get fucken WIFI from the other side of the room?

A flat-earther dies and goes to heaven. There he stares in marvel as the clouds' part, revealing that he is standing in front two magnificent golden gates. As he steps closer, he sees St. Peter and Jesus chatting and having a discussion. They stop and welcome him forward.
"Before I go in, I have to ask." says the man
"Ask anything you want, I will answer anything you want to know about the universe, space and time." says Jesus in a friendly manner.
"Was I right, is the earth actually flat?", asked the man.
"No, sorry, it is round," says Jesus
The man starts shaking his head and muttering under his breath. "It seems, this goes up higher than I thought."

Chuck Norris is the singularity

Famous Scientists

In his younger years, Copernicus was a star pupil

I was lucky enough to see a lecture from Einstein, it was great….relatively speaking

Albert Einstein is considered the father of modern physics, meanwhile his cousin Frank is the father of monsters.

Why did the chicken cross the road?

- Albert Einstein: It depends on your reference point. Did the chicken cross the road or did the road move under the chicken?
- Isaac Newton: A chicken at rest will stay at rest, until something scares it, then it will be in motion and stay in motion
- Werner Heisenberg: I can either tell you where in the area the chicken might be located or how fast it was going, I cannot do both
- Wolfgang Pauli: Are you sure there weren't two chickens occupying the same space but on either side of the road

I once met a guy who had the brain's of Einstein….. he was later arrested for grave robbing

Isaac Newton tried creating a 4th law of inertia, but it didn't get any momentum

Sir Isaac Newton was sitting under a tree wondering about gravity. Then it hit him.

Einstein and Chuck Norris got into an argument about relativity. Chuck Norris got so angry he round housed Einstein in the face. From that day forward Einstein changed his name to Stephen Hawking

How does Einstein begin a bedtime story?
Once upon a space-time that wasn't distorted by a large mass

One day Albert Einstein, while going to a press conference noticed that his limo driver looked exactly like him. A Doppelganger. After striking up a conversation, Albert decides he wants to have a little fun at this press conference and swaps clothes with the limo driver and gives him the speech.

"Don't worry, I always forget things. Just read the script and you will be fine. No-one will know." He tells the driver.

The press conference went swimmingly, and no-one expected a thing. During the end, one reporter asked a very tricky question hoping to trip up Albert. The chauffer was stuck for a second thinking, pulling at his collar.

"You know what? That is such a simple question, even my chauffer at the back of the room can answer it."

When Einstein was asked about his Nobel Prize, he said it was about time!

A bar walks into Albert Einstein and says, "oops, wrong frame of reference"

What is it called when Einstein masturbates?
A stroke of genius

On a night out Marilyn Monroe runs into Albert Einstein, they start chatting and Marilyn is her typical, flirty self when she asks, "If we were to get married our kids would be the smartest and the most beautiful in the world!"

Einstein replies, "But what if they get my looks and your brains?"

Stephen Hawking and Mr. Bean are having a chat when Stephen Hawking makes a suggestion.

"Let's play a little game. I ask you a question, if you can't answer it, I give you $1000. Now you ask me a question and if I don't answer it, I give you $1?"

Mr Bean agrees and Stephen goes first.

"Ok, what does a parsec stand for?"

Mr Bean just stands their thinking, tapping his finger against his chin, he then lets out a huge sigh, pulls out his wallet and hands him $1.

Mr Bean then asks, "Ok, what wakes up with 4 legs, walks all day on 3 legs, gets to a road and crosses it on 2 legs but at night the other legs grow back, and it has 4 legs again?"

Hawking, thinks, and thinks and is totally stumped.

"Ok, I give up." He says, "Here is your thousand dollars."

He hands the money to him.

"So, what is the thing?" Stephen asks.

Mr Bean pulls a dollar bill from the $1000 and gives it back to him.

Why was the cat so confused?
It met with Schrodinger

Why did Heisenberg hate driving?
Every time he checked his speedo he got lost

Two surfers in Hawaii just get completely wiped out by a huge wave. Both are just bobbing on their surfboards catching their breath covered in seaweed.
One of them asks. "Hey man, you, ok?"
"Yeah, I'm Feynman."

Outside a house in Wurzburg, Germany there is a sign. It reads "Heisenberg might have slept here."

Why couldn't Heisenberg perform well in the bedroom?
He never had the time, and when he did, he didn't have the energy.

Heisenberg is speeding along the back roads of Berlin when he gets stopped by the police. Furious, the police ask, "Do you know how fast you were going?"
Heisenberg says, "Not at all but I know where I am."

Neil deGrasse Tyson, Stephen Hawking and Bill Nye walk into a bar.

Neil and Bill look at each other in amazement and yell, "My God, Stephen! You're walking! You're cured!"

Heisenberg, Schrodinger, and Ohm are driving very fast down the road listening to Rammstein. When they get pulled over by the police. The officer goes to book them for speeding when Heisenberg starts arguing. There was no way the policeman could say what their speed was as he knew their exact location. The officer then starts getting suspicious and asks to check the boot. When he opened the boot, he sees a dead cat.
"Why do you have a dead cat in the boot?" the policeman asks.
"It wasn't dead until you opened it!!!" yelled Schrodinger.
The officer then decided to arrest them.
Of course, Ohm resisted.

What do you call a bunch of high-class physicists?
Feynman

Did you hear about Nikola Tesla's new car?
It was a Volts-wagon

Nikola Tesla was famous for constantly changing his mind. When asked why he couldn't just stay with one choice he said, "My thoughts are alternating currently."

It took Thomas Edison roughly 1000 attempts to make the lightbulb, he must had dark times before then.

Thomas Edison had such a charisma, he really lit up the science world.

Mina Edison was talking to her neighbour about her marriage difficulties.
"Ed just will not go down on me no matter what. We have been married for years and he still won't try it." She says all upset
"Well, dear," her neighbour replies, "George never went down on me, until I put something sweet down there, now he never stops. Why don't you give that a go?"
That afternoon she goes home, looks in the cupboard and all she has is sugar. "Damn it, why not!" she mutters to herself. She grabs the sugar and covers herself in it.
Later in the night, since the first time they were married Thomas Eddison finally performs cunnilingus on his wife and she loves it, and their marriage is saved.
Moral of the story: A poon full of sugar makes Tom Edison go down.

What do you call a stolen Tesla?
An Edison

What do Elon Musk and Thomas Edison have in common?
They both got rich off Tesla

Nikola Tesla, Osar Wilde and Charles Darwin walk into a bar. There is a punchline to this joke but unfortunately it has been patented by Thomas Edison and not able to be released to the public.

I grew up with my best friend Eddy, we were neighbours and went to high school together. I was great at science and maths, especially electricity and Eddy loved history. We spent alot of time together as best friends do. His favourite saying was "History can repeat itself!'

The years passed and we slowly both went our separate ways, I went to a university up North and studied electrical engineering. I got a job designing circuits for electric cars, it was a good job, it paid well, and they gave me a company car. Eddy got married, had 3 kids and became a devoted father. However, his son Gareth was a handful, always getting into trouble, hanging out with the wrong sort of people.

One day the company car was stolen, I reported it missing. After a day or so the police found security footage of the culprit and I raced down to see if I could identify who it was. The footage clearly showed the stolen Tesla being driven by Eddy's son.

Einstein, Newton and Pascal are together in a garden one day having a few shandies. After a few drinks they are all happy and silly and decide to play hide and go seek. Einstein goes first and closes his eyes and starts counting to 50. Pascal runs over and hides under some flower bushes. Newton goes over to stand in some dirt and draws a box under himself.
When Einstein finishes counting, he looks up and sees Newton standing there.
"Really? That was too easy, I found you." says Einstein
"Did you?" says Newton, "You found a newton over a meter square, so you found a pascal!"

A friend was visiting Niels Bohr when he notices a horseshoe hanging above his door.
"Niels, what is that? Being a man of science, I thought you wouldn't believe in superstitions."
"I don't" says Bohr, "But they say it works even if you don't believe."

Marie Curie was a brilliant scientist, but Albert Einstein was exponentially smarter than her.
$E=mc^2$

In history's greatest scientist party, all the famous ones were invited. Here are their responses to the invitations.

- Socrates: "I need to ponder over this."
- Isaac Newton: "I'll drop by."
- Charles Darwin: "I'll want to see how it evolves."
- Pierre and Marie Curie: "We're radiating with excitement."
- Georg Ohm: "How could I resist."
- Ivan Pavlov: "I'm positively delighted at the idea."
- Andre-Marie Ampere: "I will have to buy more current attire."
- Thomas Edison: "It will be very enlightening."
- Albert Einstein: "Will I move to the party, or will the party move to me?"
- Carl Friedrich Gauss: "I'm very popular at parties because of my magnetism."
- Heinrich Hertz: "I may be able to attend but I cannot say for sure how fast I will be."
- Nicola Tesla: "This sounds electrifying"

What did Nikola Tesla say after getting shocked while making a coil?
Ouch that Hertz

What is Neil's battle cry before he gets into a fit?
Prepare for deGrasse kicking!!!

What do you call it when you pour champagne all over Carl Sagan's chest?
An atro-fizzy tits

What do you call the band Nikola Tesla and Thomas Edison started?
AC/DC

One day at Caltech, a group of brilliant young physicists discover the secret to time travel. They spent months working on a machine. The way the machine worked was that they could bring, back anyone they want from the past and then put them back without making any difference in the timeline. After much discussion, the first person they decide to bring back was Einstein. They fire up the machine and in a blinding flash of light Einstein appears. After a few minutes of disorientation Einstein is astounded at the technology but wants to talk over a nice stout German ale. They take him to Walters just up the road, where everyone not only has a drink, but the place goes crazy and everyone ends up getting blind drunk with singing, dancing and people passed out all over the place. Afterwards they send Einstein back.

Then a few days later the physicists are sitting around and thinking who they should bring next. Someone yells out 'Samuel Morse'. So, with a zap and a flash of light Samuel Morse appears. After a few moments of disorientation, the group is met with disdain from Samuel. No matter what they say or ask it was met with a 'Hmmph'. They decide to take him to Walters for a drink and a feed. At Walter's he didn't touch anything at all. They sent him back.

In a discussion in the group one person says, "I don't understand Samuel, it was so confusing."

"Well," pipes in a colleague, "You can lead a Morse to Walter, but you can't make him drink,"

How does Neil cut his toenails?
Eclipse them

A young man sees Einstein on a train, "Excuse me." he asks, "but does Washington stop for this train?"

One day Bill Nye decided he had too much work to do and with that enlisted the aid of Neil deGrasse Tyson and they set about making a clone of Bill.

The cloning was a success but had 1 draw back, the clone of Bill Nye was completely evil. He burst from his cage, killed Bill, and injured Neil. Neil swore vengeance on the clone for killing his friend.

16 long years had passed, Neil had almost caught Bill's clone a few times. There was the fight in Venice, setting the library alight. Another time in Australia, an outback bar got shot to pieces causing a horrible scar on Neil's Left cheek. On this particular day in a seedy alley bar, in the back streets of Mexico city, Neil walks in, his face in a grimace and set's his eyes on the evil clone. The evil clone was lost in thought just looking down at his drink. Neil slowly walks up pulling out his Samurai sword, he won't give him a chance like he did in Bali. The clone turns to face Neil just as Neil plunges the sword into his chest.

"Why?" splatters the clone of Bill

Looking down at him Neil replies, "A Nye for a Nye."

You know there is a issue with the education system when the *Cosmos:A Space Time Odyessy* has only 1 season.

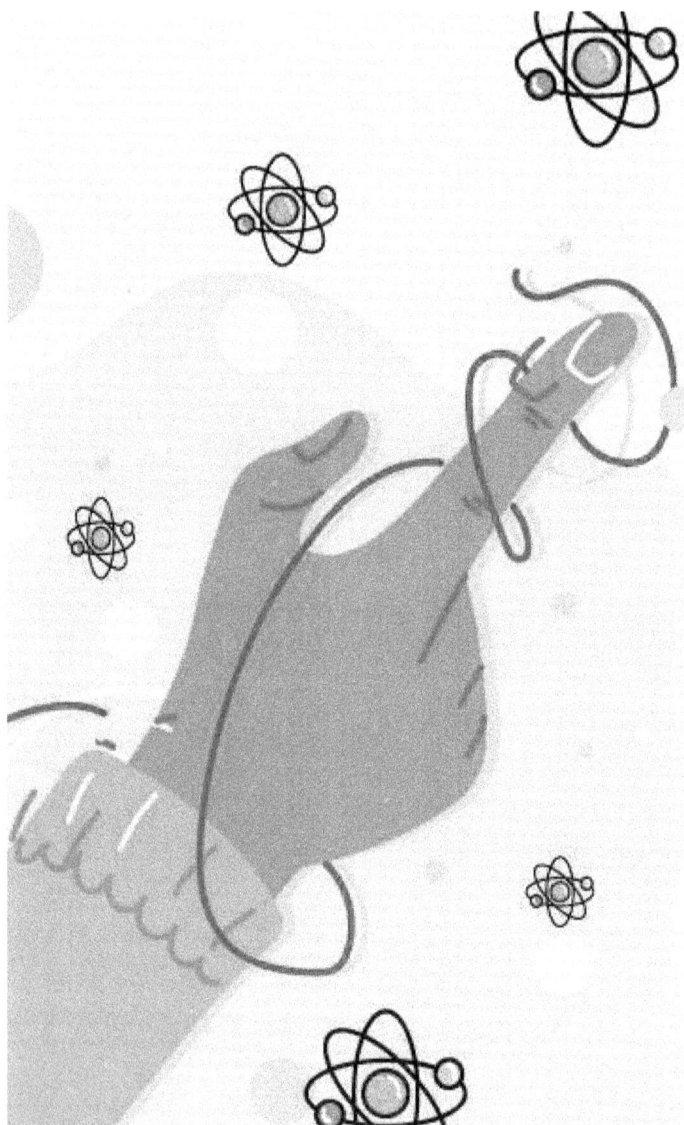

Quantum Physics

I was watching a documentary describing quantum physics. I decided not to watch the ending in case I affected the outcome

I was working on my quantum physics assignment when my mum barged in. I quickly switched it to porn. It was just easier to explain.

I caused a ruckus at the quantum physics decathlon; I was yelling out the shot-put speeds and the balls ended back at the start.

Physicists must have been breast feed to long, otherwise, why are they all obsessed with finding Higgs Bosom.

What is the difference between a quantum mechanic and a vehicle mechanic?
The quantum mechanic can get the car into the garage without opening the car door.

You can call me the Higgins Bose particle because you will be screaming "Oh God" later

What did the physicist say to a duck?

Quark, Quark, Quark

I'm like a subatomic duck that gives zero quarks about your opinion

Why did Pinocchio hate physics?
It was all about string theory

What did the physicists who built the Hadron Collider get arrested for?
Being a mass murderer

Quantum physics can give me a Hadron

What type of engineer fixes the Large Hadron Collider?
A quantum mechanic

Did you hear about this massive underground science ring?
It's called the Hadron Collider

Two physicists were talking about their recent trips overseas.
"So, I met this amazing girl at CERN, the big tunnel thingy."
"Supercollider?"

"What? NO! we only talked!"

What do serial killers and Hadron Collider engineers have in common?
They are both mass murderers

An atom went to a party at the Large Hadron Collider. He didn't recognise anyone, so he split

What did the Higgins Bose say when it was kicked out of church?
You can't have mass without me

What has no orientation and lives in the ocean?
Mobius Dick
Why did the chicken try to cross the Mobius strip?
To get to the same side.

A research centre at Barrow, Alaska has just announced a super conductor that works at room temperature

Just a random question, but what is entropy?
Well it isn't what it used to be

What is a physicists favourite WW2 rifle?
A Boson Nagant

What did the physicist name his penis?
The Hard-on Collider

A particle physicist goes up to an astrophysicist.
"Hey, do you want to hear a joke?"
"Sure", says the astrophysicist
"String theory."

What did the quantum mechanic pirate say to the classical physicists?
Walk the Planck

Relativity and Classical

Did you know that Einstein was real?
Turns out a theoretical physicist is an actual job

May the mass times acceleration be with you.

So, it turns out that the local library does have books of theorical physics. In the non-fiction section!

Friction is such a drag

Two cats are sliding off a roof. Which one falls first?
The one with the smallest mew

What happens when you cross a mountain climber and mosquito?
Trick question, you can't. A mosquito is a vector, and a mountain climber is a scalar

What band do vectors always listen to?
One Direction
People think I am lazy but I am just overflowing with potential energy

When an aircraft takes off, does that make it an incline plane?

A physicist asks his colleague, "Hey what have you heard?"
"C over lambda"

Einstein's lesser-known theory of relatives, states that the probability of your mother in-law visiting is directly proportional to not wanting to see her

What is the formula for the energy of a dinosaur?
Velociraptor = Disraptor / Timeraptor

Home is where your displacement is zero

Speed and Velocity are related, and they grew up in the same household.
Velocity was focused, worked hard, became rich and travelled the world.
Speed still lives with his parents. They say he lacks direction.

What did the physicist say when the concert was too loud?
That Hertz my ears

What are you, a ramp?
You are just inclined to agree on most matters.

What do you call the scientists who studies the gas laws by observing creaming soda?
A fizzie-cist

What people don't understand about Pontius Pilot was that he was also a physicist.
When he saw Jesus he thought there wasn't any potential in him.

Did you hear about the theoretical physicist that got bitten by a zombie?
He goes around saying "Branes, Branes, Branes"

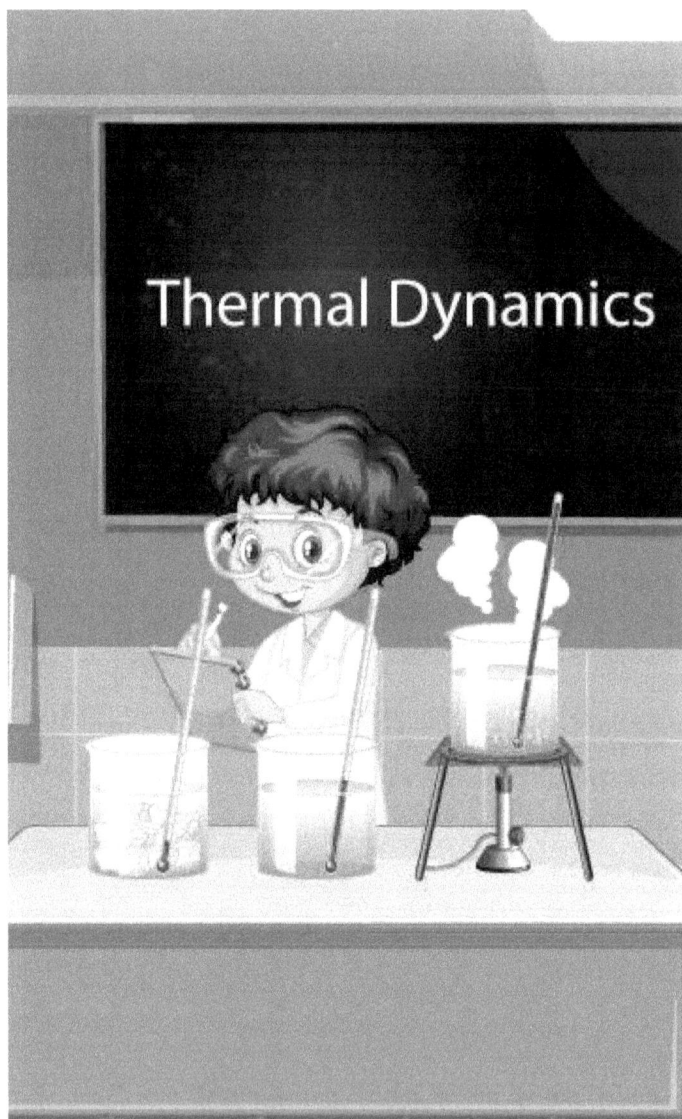

Thermal Dynamics

Thermal Dynamics

A man was chilled to absolute zero, turns out he was 0k

Don't worry about failing, absolute zero is the coolest

I learnt today that things expand when you add heat. So, I am not fat, I am just hot!

According to the second law of thermal dynamics you need to share you hotness with me

The thermodynamics of water is amazing, just look at how cool ice is

I found physics really easy but I am not doing so hot in thermodynamics

If you can apply enough heat and pressure to Kid Rock, he turns into Neil Diamond

The 3 laws of thermodynamics simplified;
- Law 1: You can't win
- Law 2: You can't break even
- Law 3: You are constantly playing

According to the Laws of Thermodynamics, consumption of alcohol in public places is an offence. Thermodynamics seems like a boring town to visit

A physicist, a philosopher and a psychologist were hunting in the woods. The three of them get lost and start wondering around the forest when they come across a cabin. They start knocking on the door asking for help, but no-one is home. Being cold, wet and hungry they decide to see if they could get in and sure enough the door was open and hesitantly entered.

It was nice and warm in the cabin with a perfect layout of lounge room, toilet and 2 bedrooms filled with stuffed animal decorations. The owner was a clearly a gifted trapper. The only thing that seemed out of place was the cast iron stove. It was an old-fashioned fire one, but it was hung off the ceiling rafters with wires. It was at least halfway up the wall.

"That's amazing" said the physicist, "he understands the laws of thermodynamics perfectly, his raised it so the heat could radiate through the room evenly."

"I don't think so," says the psychologist, "He is a lonely trapper and desperately seeking human company, without realising it, he has recreated his mother's womb."

"That's just nonsense!" says the philosopher, "Obviously he has hung up there for spiritual and religious reasons, he needs to be uplifted."

They argue about this for hours, until the trapper comes home. They immediately jump at him and ask about the stove.

"Hmm," he says grumpily, "Not much stove pipe, plenty of wire."

During university finals for physics, the very last question given to the students was:

'Is Hell Exothermic or Endothermic. Discuss using theory.'

Most students spoke of the Molecular Theory of Particles in an excited state; when they have more energy, they expand and become gas or lose energy and become liquids, then solids.

One student took a different approach:

The first constant that is needed is mass, the actual mass of hell. This, unfortunately, is not constant as hell is not a closed system; souls are entering and leaving all the time. Hell being hell would not allow many souls to escape, so it is safe to assume that the mass is constantly increasing.

How many souls are actual entering hell? If we consider the world's religions, the vast majority of them believe in hell so that those souls would go to hell. If we also look at the world's population. It is expanding in an exponential growth. The rate of births is exceeding the rate of deaths. Thus, we can hypothesise that the rate of souls entering hell would also increasing exponentially as well.

If you look at Boyles Law and the relationships between volume, pressure and temperature, we have to have to decide that either;

1. Hell's volume is not increasing faster than the number of souls entering; then, hell must be exothermic. It will keep getting hotter until all hell breaks loose.
2. Hell's volume is increasing faster than the souls entering it, which would mean an overall energy loss, then hell would be endothermic, and hell would indeed freeze over.

Which is it then?

A fellow student, Diana, had postulated to me in high school that she will only go on a date with me if hell freezes over. This was said to me 4 years ago. However, 2 nights ago, that date did occur. So now we definitely know that hell is endothermic.

On a side note, the existence of God was also proven as by the end of the night, she kept screaming, "Oh God, Oh God, Oh God!"

He received an A+

What's is the internal temperature of a Tauntuan?
Lukewarm

What is the best place to measure temperature?
K-F-C

I started a new trend. When people ask me how hot it is, I answer in Kelvins. I am slowly losing my friends by degrees

What is the temperature of a janitor's cupboard?
Broom temperature

Did you hear about the American temperature doctor?

His degree was in Fahrenheit

What is the average temperature of a coffin?
6 below

How do you take a cow's temperature?
With a thermooeter

A co-worker named Celsius recently retired. They hired another guy, Kelvin, to replace him. He's the new temp

You know what really makes my blood boil?
Temperatures above 100^0C

I accidently combined fahrenheit and millilitres today.
FmL

The Atom

What is an angler's favorite sort of energy?
Fission

Do cats around Fukushima have 18 half-lives?

What is the Incredible Hulk's pick-up line?
Do you want to feel a gamma ray burst?

What is the most terrifying sentence in nuclear physics?
That's not supposed to happen

What did uranium-235 say to lead?
I got to split.

What does a nuclear physicists have for dinner?
Fission chips

Did you hear about the latest class of see-through robots that computes physics?
It's called the new clear physicists

Why do physicists love going to church?
It is a centre of mass

A nuclear physicist goes to Oktoberfest and goes to the nearest beer tent. The barman asks, what he would like to drink.

"Ein Stein" he says

In the latest news story, a neutron was arrested under suspicion, he was later released without a charge.

Why does mincemeat provide the smallest amount of energy?
It's in its ground state

How does a physicist count sheep when trying to sleep?
1 Lambda, 2 Lambda, 3 Lamdbda

Two neutrinos walk into a bar then continue to pass through.

Why did the physicists try to stick a vacuum up his ass?
He was trying to create a farticle accelerator

You matter!!! Unless you multiply yourself by the speed of light squared. Then you are energy

Two atom's were getting ready for a fancy-dress party and trying on different outfits. One atom asks the other one.

"Do these protons make mass look big?"

Why can't you trust an atom?
They make up everything

There were once 2 physicists arguing about sub-atomic particles, they were mass-debating

What is the first song in a physicists play list?
Atomic

What do you call an atom when it dies?
A Diatom

What did the physicist say when he got into a fight?
Let me atom

Your mumma is so fat that fluorine wouldn't even react with her

What happens when electrons lose energy?
They become Bohr'ed

Hear about the 100 kilo pascals?
They got into a bar

What is a protons clothes size?
All the plus sizes

There was this child sitting on the side of the road crying when a policeman came walking past.
"What is the matter?" asked the policeman.
The child's answer consisted of weight and volume that occupies space.

Randy Scientists

A couple of scientists had twins, one was called Jessica and the other Control

Hear about the physicist that was asked on a date?
He Lepton it

I nearly had a threesome with two undergrad physicists, but they couldn't solve the three-body problem

A physicist and a biologist got married but got a divorce very early, turns out the chemistry wasn't right

Love is all about the chemistry,
but sex is all about the physics

What do you call it when a chemist gets fucked by a physicist?
Chemistry in motion

Why did the physicist like participating in threesomes?
They were interested in investigating entanglement and the double slit experiment

A physicist and a doctor both fall in love with the same woman. Every day the doctor gives her compliments and a bunch of roses. The physicist just gives her an apple. After a while the woman askes the physicists. "I understand a gift of roses, but why do you keep giving me an apple?"
"Haven't you heard the saying? An apple a day keeps the doctor away."

My ex wasn't very stable, she spontaneously decayed last week and left me for neutrino

How do people who are out clubbing define physics?
The bigger they are the easier they are to pick up

The person who I had the one-night stand with was angry with me this morning
I had to explain, I didn't have a PhD. in theoretical physics. Theoretically I had a PhD. in physics.

Why did the girl break up with her physicist partner?
Because he had no energy

Why did they patch things up?
Because he showed he had potential

Which scientists make bad lovers?
Quantum physicists, when they have the position, they don't have the momentum but when they have the momentum, they do not have the position

Astronomer: A rogue planet is one that is untethered to any star.
Relationship Expert: by the sound of it she just needs to float around and be on her own for a while and there is nothing wrong with that.

A physicist, a mathematician and a computer scientist discuss what is better. Either having a girlfriend or having a wife.
The physicists says, "I believe that having a girlfriend would be best, you still get freedom and time to conduct experiments into all hours of the night."
The mathematician, "I believe that having a wife is better, you get the security."
The computer scientists says, "I believe that having both a girlfriend and a wife is the best thing. When I am not with the wife, she thinks I am at my girlfriends. When I am not with my girlfriend, she thinks I am with my wife, but I am just playing video games all day long."

Biologists do it with clones
Botanists do it in the bushes
Marine biologists do it in the ocean
Zoologists do it with animals
Chemists do it with small balls
Physicists theorize about it.

Where do physicists go for a fun night?
The Mobius strip club

What does physics and incest have in common?
It's all about the relative

Newtons 5th Law: The intelligence of boys in an exam proportionally decreases as the number of female students increase in the same exam

Two jocks were sitting at a bar having a chat.
"How is your girlfriend going? The physics major?"
"Not good buddy, we are over, she was cheating on me."
"What? No way." says the friend
"Yeah man, I called her the other night and she said she was in bed wrestling with 3 unknowns."

What is the drive behind reproduction?
The family joules

Why can physicists only marry in deep south or middle east?
Any romantic activity is always relative.

The physics departments are getting so popular, an infra-red-light district has even opened up next door.

Pick Up Lines

I am so attracted to you, Physicists will have to name a 5th fundamental force

A chemist tells you that you are made from 60% oxygen, a biologist says that you're made of 70% water and a physicists say you are 99.99% space, but I'll say you are 100% cute

Man, I had a freak of a lab accident, it left me with a 12-inch penis

Want to test the spring constant on my bed?

I have a waterbed; do you want to come help research our densities together?

In my bed you will discover the meaning of perpetual motion

I'm hung like a Foucault pendulum

Can I have your quantum number?
What are do you prefer? A top quark or a bottom quark?

Wow you must be a third generation down quark, because you have a great bottom

So according to the Uncertainty Principle of Quantum Mechanics, you are actually in love with me, you just don't know it yet

You must be a Higgins Bose particle, because you make up my universe

They call me the Higgin Bose particle, because I'll be making you scream "Oh God" at the end

I believe that there are an infinite number of universes, that way we would definitely be together in one of them

Whatever temperature scale you use, you're still smoking hot

You are more special than relativity
Hey if your beauty was sunlight, it would still be just as bright on Pluto

I can prove to you that the Big Bang isn't just a theory

I am attracted to you like the earth is attracted to the sun-or inversely proportional to the square of the distance

I want to use my rocket ship and bombard your singularity until you go supernova

Do you want to see what's under my Kuiper Belt?

Want to couple our equations tonight?

Can I have your significant digits?

I have a tattoo on my butt saying $E=mc^2$ want to see it?

Is that a Van Der Waals force of attraction I feel or are you just happy to see me?
Babes you must be an electron, you have some potential

If your vagina was an electron, my dick would be a positron, cause I would annihilate you

I have a major in physics which means I'll never Bohr you in the bedroom

It's all about how the magnitude of the force is applied to a vector, not its actual length

We should have sex because I can really put your inertia in motion

Want to exchange fermions?

Your lab bench or mine?

You must have an phenomenal heat capacity because you are hot

Your ass is hotter than hydrogen plasma

You have the most crystal, wavelength of 450-495nm eyes I have ever seen

Want to measure the coefficient of static friction together?

You're an electric charge, and I am a magnetic charge so why don't we flux

If you were a battery, I'll be aluminium foil and together we can light things up

Are you playing hard to get? Because my voltage and your resistance will make one hell of a current

If you were a laser you would be set to stunning

Wow, you look like my supersymmetric partner

I am able to excite you to your natural frequency

That dress would look even better accelerating towards my bedroom floor at 9.8 ms^{-2}
There is a lot of potential energy between us, why don't we convert that to kinetic energy

If we got together, it would be called The Grand Unified Theory

Your love is very entropic, it increases over time just gets more chaotic

If we get together it would be like a superposition of two waves in phase

I am an astronomer, and I would love to study your heavenly body

Hey babe, how do you feel about group experiments?

Hey babe, are your related to Tesla?
because you are electrifying.

Hey babe, are you a wire conducting electricity?
Because my heart generates a magnetic field and it's
surrounding you
Hey babe, are you related to Isaac Newton?
Because even without gravity I would still fall you

Turns out Copernicus was wrong, you are the centre of
universe

I don't need a transducer to know you're hot!!

Hey babe, are you a subatomic particle?
Because I feel a strong attraction force between us

Hey babe, are you a fan of Archimedes?
Because you are about to grab my lever and shift my
center of gravity

Hey babe, have you burnt yourself?
Because when you fell from heaven your atmospheric re-
entry would have caused a massive fireball

Hey babe, are you a singularity?
Because you are so attractive, that you keep pulling me in

Hey babe, are you a singularity?
Because you are so attractive, that the closer, I get the more time slows

Hey babe, are you a singularity?
You are so attractive, you are causing a spaghettification in my pants

Hey babe, are you a singularity?
Because you have a big black hole

Hey babe, do want to see something massive?
No, Why? Don't you want to visit CERN?

Hey babe, did you just eat a magnet?
Because you are quite attractive

Hey babe, did you just eat a magnet?
Because it seems you created a strong magnetic field and induced a flow in my pants

Hey babe, are you free tonight?
Are you up for some high-energy quantum tunnelling?

What do you get when you cross the moon and the stars?
You

Hey babe, your mouth reminds me of the universe, when you are using it, it's an ideal vacuum

Hey babe, are you a physics paper?
Because I've been staring at you, and I've not understood a thing.

Hey babe, have you experienced relativistic corrections to energy levels?
Because you've got some fine structure

Hey babe, are you a quantum infinite potential?
Cause I am particle who is about to penetrate your classical forbidden region

Hey babe, have you heard of Newtons law of universal attraction?
It means your attracted to me, because I am definitely attracted to you

Hey babe, is your refractive index greater than 2.42?
Because your eyes sparkle more than diamonds

Hey babe, are you a centripetal force?
You are making my world spin

Wrong Schooling

In Physics the more you know the less you know. You start with classical physics, where you can't solve the three-bodied problem. Next, you learn the theory of relativity, where you can't solve the two-body problem. Then there's Quantum mechanics where you can't solve the one-body problem and finally, in quantum electro dynamics, you don't even understand vacuums anymore.

I could sit here all day and explain quantum physics, but you just wouldn't understand a thing. It's not that you're dumb, it's just I am a really crappy teacher.

Two physicists' students are studying for a final exam, one was quinervous so the other says to him, Don't worry about it, you are gonna be Feynman"

Young Jimmy comes home from school and his mom asks him, "What did you learn at school today?" "Today we learned about electricity, ohm my gawd it was awesome!"

Why are physics books miserable?
They are full of problems

A student asks his teacher, "What is the unit to express joules per second?"

"Yes," the teacher replies.

A physics teacher was dealing with a student who was not only the heaviest but also the rudest brat he ever came across. The student would make fart noises and steal other people's books and pens. In the last lesson the kid leaned forward and unclipped a bra strap of the girl in front.

"That's it, I need to speak to you outside!" yells the physics teacher.

"Yeah, no worries" the fat kid replies with a smirk. Outside, the teacher has now lost all his anger and seems sincere.

"Look, buddy, I honestly think you have the most potential in the class, I'll show you."

He then throws the student of the balcony

A student in a physics lecture, just before the class finishes ask, "So what happened before the Big Bang?"

"No time" answers the professor

There are three students, one is studying physics, another mathematics and the last one engineering. They are each given a ball and told to find the volume.

The maths student measures its diameter and from there works out the volume.

The physics student immerses it in water to find the volume. The engineering student starts spinning it around like crazy trying to find the model number.

Where should a physics class be held?
On the edge of a cliff. It's where they have the most potential.

In class, I got stuck on a question about friction. My friend refused to help as he is as stubborn as a μ

Why couldn't the physicist teachers get along?
There was too much friction between them

A physics exam had the following question on it.
'A young child weighing 15 kg is being held by a parent whose arms are 102cm long. The child is swinging at a revolution 0.75 cm/s and the parent lets go.
How far could the child potentially travel.'
The best answer written was 'In foster care.'

If you wake up in a science lecture, how do you know which class you're in?

- If it moves, it's biology

- if it stinks, it's chemistry
- if it doesn't work, it's physics.

A physics teacher once told me that the only way I would pass, was if pigs start flying. I studied really hard and developed my own experiment, for which I got expelled. I believe I was set up since it was never specified that the pigs had to fly on their own accord.

My favourite topic about physics was displacement, it was straight to the point.

A student kept annoying his physics teacher with questions about gravity. He was told to drop it

Little Jimmy is at school in class just watching the clock, counting down the minutes for lunch. The teacher turns around and says, "Ok class, I will give you 3 quotes, if you get one you can get who said it than you can go early to lunch."
"Who said, 'We shall fight them on the beaches?' "
Little Jimmy is there straining and thinking, when someone at the back of the room yells out, "Winston Churchill!"
"Very good, you may go." she says to the student.

"Now who said, 'Imagination is everything. It is the preview of life's coming attractions.' "

Once again Little Jimmy is wracking his brain when across the other side of the room someone yells out.

"Miss, that's Einstein"

"Very good, you can go to early lunch."

"Ok class, now who said. 'I have a dream!' "

Immediately someone yells excitedly, "Miss, Miss, Martin Luther King, Miss, Miss."

"Yes, you can go."

Little Jimmy, all angrily says, "Them bitches need to Shut Up!!"

"Who said that?!", said the teacher turning around.

"Oh, Miss. It's Bill Cosby, can I go now?"

Physics Professor: "Ok class, is there anything else you wish to know before tomorrow's exam?"

Student: "Can you go over terminal velocity?"

Physics Professor: "No."

A student asks their physics teacher if they may go to the bathroom.

"Sure," he says, "liquid solid or gas?"

How can you tell that Joe Biden failed physics?

He has the power, and he has the time, but he never gets any work done

During a bonding session, a son is asking his father all sorts of questions, covering almost every topic imaginable.

"Dad, what is string theory?", the young man asks

"Why are you asking me such difficult questions, can't you just ask me something a lot easier?" The father replies.

"Ok, so dad why is mum always yelling at you?"

"Ok Son, you see string theory is a frame work from particle physics……"

A student asked his astronomy professor, "Miss how do stars die?"

"Generally, with a drug overdose." she replied.

An undergrad student was caught cheating in his physics exam. The professor was busy giving him a lecture.

"Do you realise the gravity of this situation?" He snapped.

"That's why I cheated, because I didn't!!!"

When Hamilton applies the principal of least action he wins a Nobel Prize, when I try, I fail my exam

A physics teacher asked a student. "Do you understand Linear motion?"

"It's really straight forward," he replies.

In a school, the agriculture department needed to build a fence around some sheep. A manual arts teacher, science teacher and the senior physicis teacher go to the plot to map out the area. The manual arts teacher just looks at the material and builds a simple square fence without any fuss.

The science teacher starts pulling it down.

"No, no, no," he says "I know how to maximize the space for the material."

So the science teacher starts building a circlar fence.

The physics teacher pipes in, "Let me show how to do it."

He then grabs some material and builds a tiny fence around himself.

"See, now I will quantify myself as the outside."

In science class I hated the energy topic, it was nothing but work.

A lecturer is busy telling his physics students the philosophy of physics.

"So, if you think about it, maths is to physics, as what masturbation is to sex."

"Um, what you're saying is that without the attempt to explain the why of the universe, the search for deeper mathematical truth is lonely and pointless?" asks a student.

"No, what I am saying is that you will spend most of your undergrad just doing maths."

I failed science at school, so, to get back at my teachers I named my son Physics. Now, when the school calls me, they say. "Is this Tom, the father of Physics?"

Stereotypes

What are physicists' favourite proofs?
Half a percent of alcohol

I would always get a B in biology, a C in chemistry but I always got an F for physics

What type of beer do physicists drink?
Ein stein

The egghead scientist didn't do well at his first job, he just cracked under the pressure

A physics student asked the lecturer, "Can you help me? I need to get into a class where they teach both quantum mechanics and where I can be united with general relativity?"
"Sure let me pull some strings for you."

I love physics, but I just suck at maths, I hope it doesn't matter

A vegan physicists favourite saying is, "Now lettuce consider."

A woman goes to a doctor who tells her she only has 6 months to live. The doctor then advises her to marry a physicist and move to Martin City.

"Will that cure me doctor?" she asks.

"Sadly no, but it will make 6 months feel like 6 years."

A biologist, a physicist and a chemist go to the ocean for the first time.

The biologist is amazed and wants to see the seaweed fields and study its flora and fauna. Running straight into the ocean the rip tide promptly takes him away and he drowns.

The physicist was mesmerized by the motion of the waves. Wanting to know more about fluid dynamics and wave motion, the physicist walked straight into the ocean and also drowned.

The chemist took his notebook out and wrote: *Biologists and physicists are both water-soluble*

A physicist, chemist and a statistician are walking through the research department, when they see a trash can on fire.

The physicist goes to grab it, "We need to put this in the cool, that way the temperature will be below the ignition point and it will put itself out."

The chemist grabs his arm, "No, we just need to smother it and thus remove any oxygen. In the absence of reactants, the fire can no longer continue."
Meanwhile, the statistician starts running down the foyer, lighting all the bins on fire.
"But first we need a proper sample size."

What is the stage name of a physicists who also raps?
MC^2

The dean of a university was complaining about his budget.
"God dam physicists, do they know how much money they spend? Why can't they be more like the maths department! All they spend money on is paper, pencils, erasers, and bins. Or better yet, the philosophy department it's just paper and pens with those guys."

The mafia kidnaps a vet, an engineer and an physicist. The head mafioso makes a demand of them.
"Ok, I kidnapped you guys because I wanted a surefire way to rig horse races. If you come up with a way, I will pay you half of what you win. If not, we can't have you talking about this, so we will just put a bullet in your head."

After a while the vet goes says, "Ok, I got it. I have a list of steroids and drugs you can give the horses. It will be sure to win."

"Are you serious?" says the mafioso, "You don't think we haven't tried this before?"

He then pulls out a gun and shoots the vet.

A little while later the engineer says, "Ok I have an idea. How about we put an electric zapper on the end of the whip, so every time the jockey hits the horse it gets a little extra jolt?"

"Wow, another unoriginal idea," says the Mafioso and pulls out his gun and shots the engineer.

The physicist just looks at this and starts scribbling madly on his notepad. The next morning the physicist yells out,

 "Ok, I have it, I fool-proof way to win every race!!!"

"Ok let me hear your idea." says the mafioso

"Well, first you think of friction and gravity as negligible……"

What is a physicist's favourite fruit?
Fig Newtons

What is actually weirder than physics?
Physicists

Why do physicists love going to sport stadiums?
To study the wave

What is a physicist's favourite movie genre?
Ψ Φ

What is the difference between a physicist and a physician?
The physicist is busy before building a catapult, the physician is busy afterwards

An engineer, a physicist, a mathematician and a philosopher are at a café.
The physicists says, "You know what? Engineering is really just applied physics." and they all laugh.
The mathematician says, "Well you know what? physics is just applied mathematics." and they all laughed again.
"Well, you know what?" chimes in the philosopher, "maths is really just applied philosophy."
"Shut-up and hurry with our coffees!" says the engineer.

What does a witch and a physicist have in common?
They can make motions with potions.

A chemist, physicist and a statistician go hunting in the woods. This giant buck leaps in front of them. The chemist quickly gets off a pot shot and misses by exactly 3 ft to the left. The physicist gets on one knee, takes careful aim, and just as he shoots, he slips and misses by exactly 3 ft to the right.
The statistician screams, "Eureka, we got him!"

A dietician was asked, what is a light year?
The answer: Same as a regular year but with less calories

What do you call a female physicist that decides to explore her bisexuality?
The double slit experiment.

An engineer, a physicist and a mathematician go on a fishing charter. While on charter, there is a terrible storm and they get stranded on a deserted island.
The engineer looks around, picks up some driftwood, some flat stones and makes an axe, where he can break up the coconuts and survive. He then walks around the island looking for other survivors.

Eventually he comes across the physicist. He is busy using a stick and writing a formula down in the sand. He then walks over to a tree, studies it, then lightly hits in a certain spot before running away. A slow vibration begins in the tree and in a minute all the coconuts are falling and cracking open on the ground. The engineer and the physicist join forces and go off in search for the mathematician.

Sure enough the next they find him starving running up and down the beach. He is busy writing stuff in the sand and then running around in circles.

"I've nearly got it, I've nearly got it!" says the mathematician

"What have you nearly got?" asks the physicist

"Sssh," he says angrily, "I've nearly proved that the coconuts do not exist!"

The accounting major asks, "How much will it cost."
The physics major asks, "How does that work?"
The engineer asks, "How am I going to build that?"
The philosopher asks, "Would you like fries with that?"

What is funnier than a physics joke?
A meta-physics joke

A physicist, a scholar, an engineer and a psychic were asked what they think that the greatest invention ever was.

The physicists says that it's fire, it made man master over matter

The scholar says 'the alphabet' as it allowed humanity to record their knowledge and how to build on it.

The engineer says, it was the wheel, as it allowed man to master the space around him.

The psychic says 'a thermos flask' because it keeps drinks cold in summer but hot in winter. When the others in the group ask why that was so it important, the psychic answers, "Yes but how does it know?"

A physicist, biologist and geologist are trying to run away from fascist Germany in 1945. They are captured and then sentenced to be executed for treason to the state. As they line up and the firing squad takes position, the physicist yells, "Lightening!" All the soldiers look up, but there is no lightening and the physicist escapes.

The biologist looks around thinking, "um um, Wolves!" he yells. All the soldiers spin around, guns ready, but there weren't any wolves and the biologist escapes.

The geologist is stumped and starts thinking 'boulders, no, rocks, not metaphoric rocks, heat. That's it!'

"Fire!!!" he yells.

A farmer, philosopher and a physicist were walking along a country road. When the philosopher picks a piece of fruit off a tree, turns to the others and says, "We can never truly know what this piece of fruit is. We can make assumptions based on our experiences and wisdom plus what form this perception most fits, but we can never truly know."

The physicist takes it from his hand, looks at it and says, "Well this piece of fruit is really the result of a collection of particles colliding over and over again via fundamental forces."

The farmer looks at the pair of them. "It's a fucken orange!!"

What does the band Smash Mouth do in physics class?
Sum Bodies

Two mathematicians and Two physicists are going to a convention across the state. They decide to catch the train. At the station the physicists buy two tickets, but the mathematicians only buy 1. The physicists just look at the mathematicians with a funny look but don't mention anything.

As the train ride progresses the physicists see that a conductor is going up the train checking for tickets.

"Hey guys, the conductor is checking tickets." they warn the mathematicians.

The pair of mathematicians quickly get up and go into the bathroom and wait.

The conductor comes up to the door, knocks and yells, "TICKET."

A train ticket slides under the door, the conductor takes it and walks off. A few minutes later the pair come out.

"That is brilliant, we will have to try that on the way home," says one of the physicists.

On the train ride home, the two physicists buy only one ticket but this time the mathematicians don't buy a ticket at all. Sure, enough the conductor comes back through, looking at tickets. The physicists quickly get up and run to the bathroom. A few seconds later the mathematicians walk up to the bathroom door, knock and yell "Ticket." The physicists then slides the tickets under the door. The mathematicians grab it and run into the next carriage toilet.

"Typical physicists, they use all our concepts without truly understanding the workings," says one of the Mathematicians

Glossary

A

Absolute Zero: so cold that it will actually freeze your nuts off, even if you don't have any.

Absorb: what happens when someone's attention is focused on one particular object, like when a physicist finally steps out of a lab and is amazed by how bright everything is.

Acceleration: what happens when you're driving and a spider runs across your hand, and your foot slams on the wrong pedal.

Accuracy: how close to the centre all your shots were to the target. A high accuracy doesn't mean you're any good.

Adhesion: how strong a glue can be. I once used a super glue that was that good that I couldn't remove a plank of wood from my face for 3 days.

Advection: an advertisement for vection.

Alhazen: short for Abu-ali al-hasan ibn al-Hasan ibn al-Haytham. He proved that sight occurs due to external rays, nearly a thousand years before it was 'discovered' by the West

Ampere, Andre: made maths hard for high school children for hundreds of generations due to his work in partial differential equations. Also, Amp was named in his honour

Amplitude: an instrument used to make speakers louder

Anode: where the electrons are running away from

Archimedes: someone who spent too much time in a bathtub

Arrhenius, Svante : figured out that the hotter things are, the quicker something cooks

Atomic Mass : where atoms go to pray

Atomic Mass unit: when all religious atoms meet and have a discussion

Atomic Orbitals: mini solar systems

Attenuation: the only part of physics, that surfers are actually interested in

B

Bacon, Francis: argued that science needed evidence. Unfortunately, this still needs to be explained to people

Beam: cars have 2 different modes of beams, and some wankers never seem to dim theirs.

Bernoulli, Daniel: analysed water moving through a hole, perhaps a golden shower fetish

Big Bang: when everything was created, in a loud explosion. It also begs the question, 'Did it make a noise, since no one was there to hear it?'

Big Bang: when a physicist loses his virginity, usually at the age of 23.

Black-bodied radiation: a racist radiation

Bohr, Niels: famous physicists, form whom all chemists stole and manipulated his ideas.

Boyles, Robert: some dude who used to fart in bottles, he did figure out pressure volume relationships

Brown, Robert: observing pollen granules in water and drunk people hailing taxis, Robert came up with the theory 'random walk'.

Brownian motion: how drunk people get home

Buoyant: opposite of a girl ant

C

Carnot Cycle: a mythical cycle that technically cannot exist

Carnot, Sadi: someone who ignored entropy to make the maths easier

Cathode: what the electrons are drawn to

Centrifugation: the best ride at the fairground

Chadwick, James: loved bombarding tiny things with smaller things

Charge: you get one if you are a naughty physicist

Clausius, Rudolf: had a shiny nose and saved Christmas one year.

Combustion: what happens when a large-breasted woman wears a very tight shirt.

Conductivity: what train drivers do

Coriolis effect: if movies have taught me anything, snipers use this knowledge all the time

Curie, Marie: a women who won, Nobel prize, in radiation. Played with it so much that her house is still a biohazard. She did not get super powers.

Curie, Pierre : First man in history to be completely overshadowed by his wife. Also died of radiation positioning, did not get super powers.

Current: a type of fruit often confused with a grape or berry

D
Davisson-Germer experiment: showed that electrons can move in waves, this information was then used to give other people a Nobel Prize

de Broglie, Louis: had an idea, then took Davisson-Germer data to prove it , then got a Noble Prize without sharing credit

Decay: usually where de beach is

Density: a level of how not smart someone is

Diffraction: bouncing things around, if you can work this out then you are a master at snooker

Doppler Effect: imagine sirens screaming past you, it also does not consider colour blind people

E

Effusion: farting

Einstein, Albert: famous quote, "If I had only known, I would have been a lock smith"

Einsteinium: this element defines irony, an element that was discovered by flying a military jet armed with a scoop through nuclear fall out after testing a thermonuclear device. Then this element was named after a pacifist who hated the idea of his theory being used to make atomic weapons.

Electrical Charge: your quarterly electricity bill

Electrical Charge: what electrons get taken to the police station for

Electrons: One of the most valuable resources in the world

EMIT: when an electrons confesses to the charge

Endothermic: something that sucks the life out of you, financially, spiritually and mentally

Entropy: trying to describe confusion but sounding smart at the same time

Exothermic: when you are so hot, that your sweat steams

Extrapolate: the only thing theoretical physicists actually do

F

Faraday, Michael: allowed hackers everywhere to build cages around themselves so the government and aliens can't track them

Fermi, Enrico: A Manhattan Project scientist who helped find the big elements. Did not get superpowers of gamma rays

Feynman, Richard: A Manhattan Project scientist, Investigate the challenger incident, wrote several books about himself, loved nice dinners out and fancy clothes, you could say he was a fine man. Also did not get superpowers from gamma rays.

Fick, Adolf – great scientist, bad name

First Law of Thermodynamics: is that you do not talk about thermodynamics

Force: the energy that flows through us and around us. It's strength is determined by how infested we are with a parasite called mid-chlorians.

Force: Chuck Norris

Franklin, Benjamin: in a lightning storm holding a kite attached to a key he proved everyone wrong by electrocuting himself.

G

Galileo, Galilei: took a pirate spyglass and looked up, then called the spyglass a telescope.

Gas: another name for farts (the most annoying type of jokes and the infernal "Pull my Finger")

Gay-Lussac, Joseph-Louis: the father of pull my finger

Gravity: what goes up must come down, unless it, of course reaches escape velocity.

Gutenberg, Beno: Another physicist who the other sciences (in particular geology) need to thank for solving their problems

H

Hau, Lene: So not only did she slow down light, but this space wizard changed light into matter then back into light again. Seriously where is her Nobel Prize?

Heat: how bouncy atoms are inside something

Heisenenberg, Werner: if after reading this book you do not understand his principles, I suggest you start reading this entire book again.

Higgs boson: theorized in the 60's but not discovered until 49 years later. That was a long time to wait to be awarded a Nobel Peace Prize.

I

Ideal Gas: 'pull my finger.'

Isaac Newton: please refer to Heisenberg, Werner, same meaning applies

J
Joule, James Prescott: Published a paper describing Joules Law. It states that the size of the diamond in a engagement ring is inversely proportional to the waist size of the bride to be

K
Kelvin, William: rewrote the temperature scale to make 0^0 absolute freezing and used his last name as the scale
Kinetic Energy: how hard you can get slapped

L
Light: opposite of heavy.
Line Spectra: a ghost that exists in 2D land
Liquid: a stiff drink
London dispersion force: The London police riot squad

M
Magnet Moment: the moment you meet the love of your life
Magnetism: something physicists actually have but other scientists lack

Mean free path: getting out of the bouncer's way at a club

Michell, John: played with magnets

N

Newton: A unit for measuring force, named after the Issac Newton. It correlates directly to how hard Newton slapped people. For example, 2 newtons is equal to 2 Newtons slapping you.

Non-relativistic – in physics when the maths gets too hard you can simply say it was non-relativistic and remove sections of it.

Nuclear Fission: the power of the sun. What we have been trying to achieve for the past 80 years.

Nuclear Fusion: how atomic bombs work

O

Oersted, Hans Christian: famous author of children's books. Eg. The Little Mermaid and Thumbelina

P

Pascal, Blaise: was really good at working under pressure.

Pauli exclusion principle: showed that electrons are snobs and believe in monotony

Planks Constant: describes that there will always be some idiot trying to do this on a balcony somewhere

Potential Energy: how much energy is bottled up when the partner says, "I'm fine or nothing's Wrong"

Q

Quantized: the act of placing the quantum in front of anything to make it appear smarter and more expensive

Quantum Mechanics: a team of people hired to fix quantums when they breakdown

Quantum numbers: A set of points where Quantum mechanics look for things

R

Radiation: something that does not give you superpowers

Rarefactions: a faction that is rare and determined by UNESCO

Rayleigh surface wave: named after an extremely obese person. The surface wave is what you see when you slap Rayleigh's (or an obese persons) belly

Redshift: red does go faster

Reflect: bouncing off something

Refraction: bouncing through something

Relativistic: when physicists are related

Relativity: incestuous fun

Rutherford, Ernest: co-creator of the video game Half-life

S
Schrodinger, Erwin: crazy cat killer, PETA was definitely not a fan

Second Law of Thermodynamics: with every reaction there is a increase in chaos

Soddy, Frederick: worked in physics but got an award in chemistry.

Sonar: how bats and dolphins see

Spectrogram: measurement of weight for ghosts

Spectrograph: a graph representing the exponential growth of ghosts

Spectrometer: an SI unit for measuring the size of ghosts

T
Theorem: a statement that is true and can be proven using reasoning, maths and physics. Something a non-science person doesn't understand and often followed with the statement, "It's only a theorem."

Thermal: underwear you use in extreme cold to keep your genitals warm

Third Law of Thermodynamics: technically cannot exist

U
Ultraviolent: the wavelength aliens can see in

V
Velocity: using Jeremy Clarkson's immortal words, "SPEED!!!"
Viscosity: how sticky something is

W
Wave function: a group of people getting together and celebrating the Mexican wave
Wave period: that special time for a wave
Wave-particle duality: when a wave is bisexual
Wavelength: when male waves get together and try to outdo each other by talk about the size of their penises
Waves: greeting someone
Weight: standing in line for 2 hours to renew your driver's licence
Wilkins, Maurice: worked with radar and the Manhattan Project, later turned his attention to DNA, he was the original Bruce Banner but unfortunately, no accidents happened, so we never got the Incredible Hulk.
Work: where you exchange your time for money

X
X-ray: looking at someone's insides

X-ray diffraction: how the x-rays bounce around inside of you

Y

Young, Thomas: inspiration for the hero in the mummy, both a physicists and a Egyptologist, proved that light was a wave and deciphered the Rosetta Stone

Albert Einstein misquote- "If you enjoyed this book, you'll enjoy the free inappropriate adults only joke book at

www.unconventionalpublishing.com.au "

Authors Message

Hi, there.

Thank you for taking the time to read this book. Coming from a science background I have been wanting to write these joke books for a long time and unleash the inner nerd in me. I hope you had just as many laughs as I did in writing this book. If you did enjoy reading this, please leave a review that would be greatly appreciated, and would help me out immensely.

**Free Adult Jokes Book **

You can also find another free joke book on the company's website, one that is not for sale, one that is not for the faint hearted, one that can be considered a bit risqué and politically incorrect. Consider this a gift for taking the time to purchase and read my book but be warned, only download it if you are not easily offended.

www.unconventionalpublishing.com.au

If you want to see any other professions being roasted or want a particular joke on a shirt, please let us visit our website and let us know.

Kind Regards

Shane Van

www.ingramcontent.com/pod-product-compliance
Lightning Source LLC
Chambersburg PA
CBHW051858090426
42811CB00003B/373